FANTASTIC PLANTS

PRICKLY DESERT CACTI

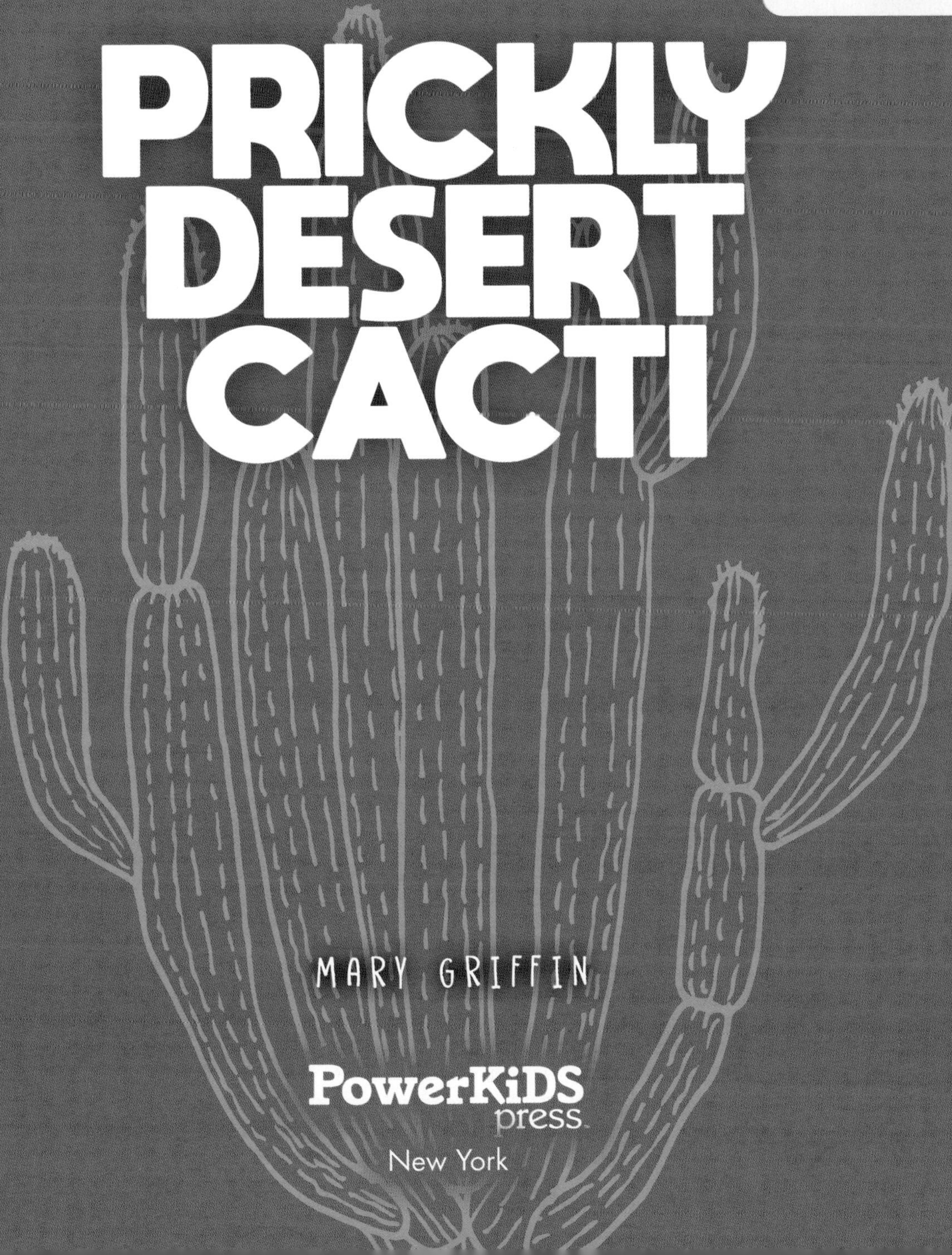

MARY GRIFFIN

PowerKiDS press
New York

Published in 2023 by The Rosen Publishing Group, Inc.
29 East 21st Street, New York, NY 10010

First Edition

Portions of this work were originally authored by Tayler Cole and published as *Cacti Barely Need Water!* All new material in this edition was authored by Mary Griffin.

Editor: Therese Shea
Book Design: Michael Flynn

Photo Credits: Cover Delaney Van/Shutterstock.com; back cover, interior background (illustration) Foxyliam/Shutterstock.com; interior frame Liubou Yasiukovich/Shutterstock.com; p. 5 BeataGFX/Shutterstock.com; p. 7 chatupol pungpa/Shutterstock.com; p. 9 Squeeb Creative/Shutterstock.com; p. 11 Alexander Seleznyov/Shutterstock.com; p. 12 Cornel Constantin/Shutterstock.com; p. 13 Mike Ver Sprill/Shutterstock.com; p. 14 Bill Florence/Shutterstock.com; p. 15 LHBLLC/Shutterstock.com; p. 17 Barbara Ash/Shutterstock.com; p. 19 (cactus illustration) bsd studio/Shutterstock.com; p. 19 (upper left) kojihirano/Shutterstock.com; p. 19 (middle left) onairp/Shutterstock.com; p. 19 (bottom left) Ken Griffiths/Shutterstock.com; p. 19 (top right and middle right) Charles T. Peden/Shutterstock.com; p. 19 (bottom right) IrinaK/Shutterstock.com; p. 21 high fliers/Shutterstock.com.

Library of Congress Cataloging-in-Publication Data

Names: Griffin, Mary, 1978- author.
Title: Prickly desert cacti / Mary Griffin.
Description: New York : PowerKids Press, [2023] | Series: Fantastic plants | Includes index.
Identifiers: LCCN 2021053822 (print) | LCCN 2021053823 (ebook) | ISBN 9781538386675 (library binding) | ISBN 9781538386651 (paperback) | ISBN 9781538386668 (set) | ISBN 9781538386682 (ebook)
Subjects: LCSH: Cactus--Juvenile literature. | Cactus--Ecology--Juvenile literature.
Classification: LCC QK495.C11 G75 2023 (print) | LCC QK495.C11 (ebook) | DDC 583/.885--dc23/eng/20211103
LC record available at https://lccn.loc.gov/2021053822
LC ebook record available at https://lccn.loc.gov/2021053823

Manufactured in the United States of America

CONTENTS

SUPER SURVIVORS

Picture a cactus in your head. You're likely thinking of a plant with thick "arms" and sharp **spines**. But nearly 2,000 species, or kinds, of cacti exist. These **succulents** range in size from less than an inch (2.5 cm) to 60 feet (18 m) high!

Cacti are native to North and South America. Mexico has the most species. Most have **adaptations** that help them live in hot, dry areas. But some **survive** on mountains, in rain forests, and even in cold parts of Alaska!

BE AN EXPERT!

THE WORD "CACTUS" MEANS JUST ONE OF THE PLANT. "CACTI" MEANS MORE THAN ONE.

CACTI ARE PERENNIAL PLANTS. PERENNIAL PLANTS LIVE SEVERAL YEARS AT LEAST.

NOT THIRSTY

When you're in the hot sun, you might get thirsty. You sweat out water, and you need to replace it. Cacti don't get thirsty. They store water well for a long time in their stems. Stems can swell to hold even more water. They're covered with a waxy skin that keeps the water from **evaporating**.

Cacti have a wide system of shallow roots to collect rainwater. Some have a long root called a taproot that grows deep into the ground. It holds the cacti in place.

BE AN EXPERT!

IN THE DESERT, RAINWATER CAN EVAPORATE FROM THE HOT SURFACE QUICKLY. A CACTUS'S ROOTS HAVE TO BE NEAR THE SURFACE TO TAKE IN THE WATER BEFORE THAT HAPPENS.

CACTI STEMS OFTEN HAVE RIBS, WHICH ARE LONG RAISED BUMPS, OR GROOVES. THESE CAN BE HARD TO SEE WHEN THE PLANT SWELLS WITH WATER.

DON'T TOUCH OR OUCH!

Most cacti don't have regular leaves. Spines are a special kind of leaf, though. Their shape keeps the plant from losing too much water through them. The spines grow in small clusters from little mounds called areoles. Depending on the species, cactus spines can be short or long, soft or sharp, and straight or hooked.

Many spines on a cactus offer shade from sun so the cactus doesn't overheat. The spines stop many animals from eating the cactus too!

BE AN EXPERT!

CACTI USE SUNLIGHT, A GAS CALLED CARBON DIOXIDE, AND WATER TO MAKE FOOD FOR THEMSELVES THROUGH A PROCESS CALLED PHOTOSYNTHESIS. THIS HAPPENS IN A CACTUS'S STEM.

AREOLES ARE ARRANGED IN PATTERNS ON A CACTUS'S STEM.

GROWING NEW CACTI

All cacti can reproduce, or make new plants. Here's one way: A part of a cactus flower called the stamen makes pollen. This pollen must reach the flower part called the pistil on a different cactus plant. If this happens, the flower that receives the pollen will produce seeds within fruit. Animals eat this fruit and spread the seeds. In the right conditions, new cacti will grow from the seeds.

Some cacti can reproduce another way. A part of the plant can break off, grow roots, and become a new plant!

BE AN EXPERT!

CROSS-POLLINATION IS THE MOVEMENT OF POLLEN FROM ONE FLOWER TO ANOTHER. THIS CAN HAPPEN THANKS TO WIND, BIRDS, BATS, SPIDERS, AND BUGS.

POLLEN LOOKS LIKE YELLOW DUST. BUGS AND BATS MOVE POLLEN FROM PLACE TO PLACE WHEN THEY'RE DRINKING **NECTAR** FROM DIFFERENT CACTUS FLOWERS.

CACTI AT NIGHT

Most plants have tiny openings, or pores, called stomata. These open up during the day to take in carbon dioxide, the gas needed for photosynthesis, from the air.

But to avoid the hot, dry desert air, cacti keep their stomata closed during the day. They open them when it's cooler at night. This is another adaptation to avoid water loss. Cacti save the carbon dioxide until daytime. Then they use it to make food for themselves through photosynthesis.

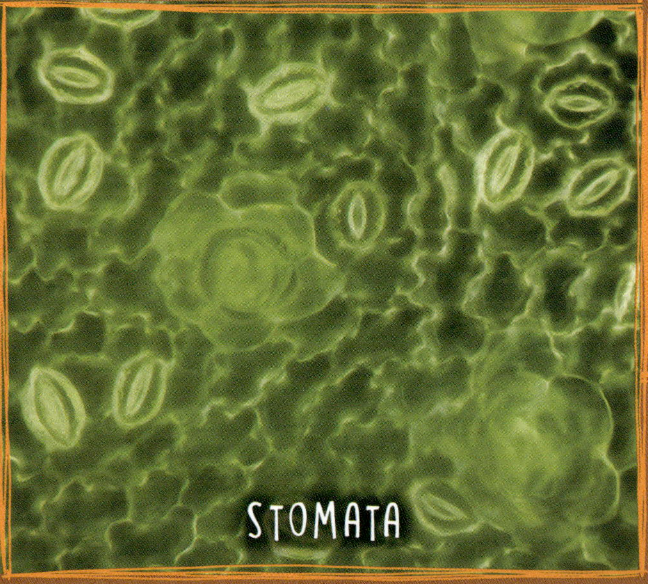
STOMATA

SINCE A CACTUS DOESN'T HAVE TRUE LEAVES, ITS STOMATA ARE IN ITS STEMS.

THE SAGUARO

The saguaro cactus is one of the most famous. Its tall, thick stem and curved arms are easy to spot. A saguaro can live up to 200 years! Saguaros are found in the Sonoran Desert of southwestern United States and northwestern Mexico.

During May and June, the saguaro grows white flowers that open at night. They only stay open until the next afternoon. These plants grow red fruit. This fruit was an important food for Native Americans in the area.

SAGUARO FLOWERS

BE AN EXPERT!

SAGUAROS GROW VERY SLOWLY AT FIRST. AFTER THEIR FIRST 10 YEARS, THEY MAY ONLY BE ABOUT 1 INCH (2.5 CM) TALL!

THE SAGUARO FLOWER IS ARIZONA'S STATE FLOWER. THE FLOWERS GROW ON TOP OF THE CACTUS'S STEM AND BRANCHES.

CACTI YOU CAN EAT!

Several species of cacti are called prickly pear, or nopal. Some kinds can grow to be 7 feet (2.1 m) tall. Their stems, called pads, are flat and branch off each other. The pads may or may not have spines. But both pads and fruit have tiny hairlike thorns called glochids.

The prickly pear has yellow, red, or purple flowers. These flowers can produce fruits called tuna or cactus pears. Like the fruits, the pads can be cooked and eaten—after the spines and glochids have been taken out.

BE AN EXPERT!

PEOPLE WEAR GLOVES WHEN THEY PICK THE FRUIT OF THE PRICKLY PEAR CACTUS. WHEN GLOCHIDS GET IN YOUR SKIN, THEY'RE HARD TO GET OUT!

PRICKLY PEAR CACTI ARE FOUND IN THE DESERTS OF THE AMERICAN SOUTHWEST. THEY'VE BEEN PLANTED IN AUSTRALIA, AFRICA, AND OTHER PLACES TOO.

HOW CACTI HELP

A cactus is an important part of its **ecosystem**. Birds use cacti branches to build nests. The spines keep their nests safe. Some woodpeckers make homes *in* saguaros!

During their **migrations**, bats drink the nectar of the saguaro and organ pipe cacti. Cactus bees use pollen to feed their larvae. Desert animals also eat cacti and their fruit. While people can eat some cacti, we can only drink from one species of barrel cactus, the fishhook cactus.

BE AN EXPERT!

MOST CACTI DON'T STORE WATER IN THE SAME FORM THAT WE DRINK. THE LIQUID IN THESE CACTI WOULD MAKE YOU SICK!

WHAT CACTI PROVIDE

stems are homes for animals

nectar is food for bats and others

pollen is food for bee larvae

pads are food for rabbits and others

fruit is food for animals and people

sources of water for animals and people

THIS PLANT IS A SOURCE OF FOOD AND **PROTECTION** FOR MANY CREATURES.

RESPECT THE CACTI

Some species of cacti are endangered, or in danger of dying out. Laws protect them. People can't dig them up for their collections, for example.

Global climate change puts cacti in danger too. Hotter temperatures can make wildfires that burn the plants even worse. And young cacti may not survive in **drought** conditions like older cacti can. Cacti are survivors, but their adaptations can't save them from every danger. People need to respect and help these amazing plants.

BE AN EXPERT!

AN **INVASIVE SPECIES** CALLED BUFFELGRASS IS A DANGER TO SAGUAROS. IT TAKES WATER FROM THE SOIL AND HELPS WILDFIRES SPREAD AND BURN HOTTER.

THE SAGUARO IS CALLED A KEYSTONE SPECIES OF THE DESERT. THAT MEANS IT'S NEEDED FOR THE HEALTH OF THE ECOSYSTEM AND EVERYTHING LIVING IN IT.

GLOSSARY

adaptation: A change in a living thing that helps it live better in its habitat.

drought: A period of time during which there is very little or no rain.

ecosystem: A natural community of living and nonliving things.

evaporate: To change from a liquid into a gas, or vapor.

global climate change: Change in Earth's weather caused by human activity.

invasive species: Plants or animals from one area that spread quickly in a new area and harm native plants and animals.

migration: The act of a group of animals moving from one location to another.

nectar: A sweet liquid made by flowering plants.

protection: Something that keeps something else safe.

spine: A stiff, sharp-pointed plant part.

succulent: A plant that is able to exist in dry places by using water stored in its leaves or stems.

survive: To keep living.

FOR MORE INFORMATION

BOOKS

Bishop, Celeste. *Plants That Poke.* New York, NY: PowerKids Press, 2017.

Johnson, Rebecca L., and Phyllis V. Saroff. *A Walk in the Desert.* Minneapolis, MN: Lerner Publications, 2021.

Miller, Mirella S. *Desert Ecosystems.* Mankato, MN: 12 Story Library, 2018.

WEBSITES

Cacti
www.nwf.org/Educational-Resources/Wildlife-Guide/Plants-and-Fungi/Cacti
Read about the smallest and largest species of cacti.

Fun Cactus Facts for Kids
www.sciencekids.co.nz/sciencefacts/plants/cactus.html
Want to know more fun facts? Read them here!

INDEX